“Even
After
All this time,
The Sun never says to the Earth,

'You owe me.'

Look
What happens
With a love like that.
It lights the whole sky.”

[Hafiz]

THE BLUE MOON

Published by Blue Moon Publications
London, England

A catalogue record of this book is available from the British Library.
ISBN 9798655722903

Cover design and type-setting by Abu Shama
Editor Ayesha Khan

Additional information and resources, please visit us on
www.bluemoon.org.uk

DEDICATION

This book is dedicated to my dear friend, who inspired me to write it. Incidentally, his name is Qamar (the Arabic word for Moon). And I am often called Shams (the Arabic word for Sun).

About the Author

Abu Shama has a Bachelors in Mechanical Engineering and a Masters in IT. He worked within management consultancy and advisory for a decade. Abu then went on to qualify as a mathematics teacher. He is passionate about learning and has a holistic approach to education. It is whilst teaching, he discovered his ability to tell captivating stories.

In this book, Abu shares his educational philosophy of combining multiple disciplines including science and art, astronomy and poetry, as well as love and climate change. Through this book, he wants to evoke the human spirit of love, kindness, beauty and hope that lies within readers.

> *"The earth is resting on a tipping point, she needs all her friends to come to her aid, lend her a hand and pull her back from the brink. If she is given a chance, she will regain balance and flourish, which she has done time and again in the past, and will continue to do so for generations to come."*
>
> [Abu Shama 2021]

The Blue Moon

Why is the moon so blue?

Abu Shama

"A million lifetimes ago…

Two celestial bodies encountered

FATE'S arrow.

A chance meeting brought them together,

A momentary embrace, then separation soon after.

As they untangled and parted ways,

Each left behind a token and a lasting gaze.

One celestial became the wondrous Earth,

Moon the other, smaller but no less in worth.

Earth rejoiced and celebrated their brief union,

Moon lamented their parting and pined for

REUNION.

Whenever Earth looked up, there he was, brilliant and bright,

Bathing the evening sky with soft, silver light.

To Earth,

Moon was a constant companion, a beacon of hope,

Radiant in all phases, never appearing as one to mope.

Yet, beyond the surface of his charm and mirth,

Moon had a sullen side, unbeknown to Earth.

Moon beheld Earth with fondness and concern,

Wherever he looked, incredible

METAMORPHOSES

transpired at every turn.

Over time, he saw

Volcanoes erupting violently, as the sky roared and rumbled,

Earth quaked wildly, as lava flowed, rolled and crumbled.

New lands formed and mountains began to soar,

Gases spewed, condensed and started to pour,

Meteorites and comets hauled down fire and ice,

A blazing inferno she was, far from the blue-green paradise.

Earth cooled, atmospheres arose and vowed to never escape.

Earth was shielded, she was terraforming and taking shape.

With time…

Moon's worry and concern gave way to happiness and joy,

For Earth staged the most spectacular show for him to enjoy.

Moon looked on in awe of her clouds, thunder, lightning and rain,

And marvelled at the multi-colours of her rainbows and auroras, time and again.

Moon smiled as the water flowed, carving streams through the land,

Rivers emerged and encountered the sea, besides shores of sand.

And then... amidst her seas,

LIFE…

came to be in slow degrees.

As time elapsed…

Simple life mastered how to harness the energy of the Sun,

Earth breathed. The great

OXIDATION

of the planet had begun.

Soon…

Life diversified and new life-forms emerged thick and fast,

Each more complex and accomplished than the last.

Plants sprouted in spring, birds sang in and amongst the trees,

Insects appeared and animals flourished on land and in the seas.

Earth now, enlivened with a thousand splendid sounds.

Layers of fragrance perfumed her from the sky to the ground.

Life draped her in a multitude of colours and the most beautiful gown,
And transformed her as the

JEWEL in the heaven's crown.

Earth looked stunning, shimmering like an exquisite pearl,
Amazing what can arise from a cloud of dust and swirl!

The sky was adorned, beholding her sight,
She filled Moon's entire horizon with delight.

Love blossomed in the cold emptiness of space,
The centre of his universe she became by love's grace.

Every day,
Earth pulled Moon at the innermost part of his core.
His longing for her stirred a fathomless sea without a shore.

Tides of emotions surged forth; rising high and falling low,
Mirrored too in Earth's oceans as tidal currents;

EBB and FLOW.

Earth had captured his heart from his initial gaze,
And his love for her enslaved him till the end of days.

Mesmerised and bewildered, he sought freedom in separation,

Yet, his longing made him a captive even in his isolation,

Torn between, coming, staying and taking leave.

Moon danced, waxing, blooming and waning in a

TIMELESS weave.

Round and round, Moon followed Earth in an endless chase,

Two celestials bound in union for all time and space.

Millions of years had now flown past…
Life on Earth had grown ever more varied and vast.

Moon witnessed the dawn of the

DINOSAURS,

Some on two legs and others on all-fours.

They roamed through Earth in the Triassic years,
And thrived in the Jurassic period without fears.

They dominated the Cretaceous era from one continent to the next,
The fiercest of them all was the mighty Tyrannosaurus-Rex.

And on one eventful summer's morning,
Fate appeared without forewarning.

The thread of fate weaved together the fabric of space once more.
Who could have known of the impending sorrow in store?

Moon held his breath in disbelief as…

WHOOSH,

A giant asteroid hurtled passed him with a whistling swoosh.

Collision was imminent, the asteroid had set course for Earth,
The heavens gasped… doesn't this brute know her worth?

A deafening

BOOM!

Followed by dread, death and doom.

Earth was struck with a mighty blow,
The impact was dealt by a brutal foe.

The tremendous force peeled the surface of her skin,
Still, the asteroid burrowed further, deeper within.

Shockwaves travelled around the world at great speed.
Wounded and defenceless, she had no time to prepare, nor to plead.

A million tons of debris spewed up from the initial blast,
The sky turned vengeful as fireballs rained thick and fast.

Firestorms encircled Earth, rapidly moving from the East to the West.
Ash clouds engulfed her; suffocating the life-breath out of her chest.

Harrowing cries echoed and reverberated across the sky,
Merging together as one final heart-rending sigh.

Then the murmur faded into the distance,
And was replaced by the sound of...

SILENCE!

The beautiful summer's morn quickly turned to night,
Earth was plunged into darkness without sunlight.

Years passed by, the Earth lost touch,
Besides her he remained, keeping vigil, standing watch.

He was a loyal companion, anxiously onlooking,
Hoping, praying and

Then…

On one spring morn, the smog lifted and withdrew,

And the utter desolation came into full view.

Moon saw in the clear light of day,

The dinosaurs that reigned supreme had been swept away.

The marine reptiles that roamed the ancient seas,

And the pterosaurs that soared upon the ocean breeze…

EXTINCT, all gone!

Now only their remnants lived on.

Earth, a shadow of her former self – withering away.

Barren, devastated and in utter disarray.

And then…

From the corner of his eyes, he noticed something through the struggle and strife.

A smile arose…

Goodness gracious, she still had signs of…

LIFE!

Life always finds a way to rebuild.

A great many millennia passed,

Earth made a full recovery at last.

In fact, she was more beautiful than ever before,

Lush meadows and pristine seas, teaming with life once more.

From the tiniest phytoplankton to the giants of the sea,

Countless plants and animals species, living as one – in harmony.

Life's rich

BIODIVERSITY

was the most wondrous spectacle to behold,

Earth's greatest triumph and tragedy, perhaps, still waiting to unfold….

ZOO

Moon witnessed the dawn of a new era with the rising of the Sun,
The Holocene age of wilderness and stability had begun.

Formed of fire, wind, air, water and the very heart of Earth,
Man sprouted forth humbly as Earth rejoiced of his

BIRTH.

Life on Earth soon settled into a gentle seasonal rhythm,
Man lived alongside nature, learning and growing in wisdom.

He sowed love within the heart of the Earth,
And she nourished Man with harvest and mirth.

Man cultivated and replenished her on a daily basis,
And in return, she turned her deserts into lush oases.

Then…

All of a sudden, Man's nature changed within a

BLINK,

Moon looked in shock and horror causing his heart to sink.

Man forgot he was a part of Earth and that 'mother-child' bond,

Standing apart from her, his eyes cast on the 'prize' beyond.

Blinded by greed and conceit – Earth became 'a thing' to conquer,

Ushering the Anthropocene age of dominance and error.

Endowed with intelligence, language, reason and imagination,
The pace of Man's progress was unlike any other creation.

Man explored and probed, traversing Earth – remote and wild,
He dissected and delved deep, his curiosity – unsatisfied.

Man mastered the way of the natural order and how to subjugate,
He utilised the seasons, tamed the wild, multiplied and then came to

DOMINATE.

Man meteorically rose as the sole ruler and centre of all.
Now 'the master', he determined the future of everything, large and small.

Soon after…

Man harnessed the elements, unlocked the genome and split the atom,

He wielded great power and potential, it was his time to lead and blossom!

Instead…

Conflicts ensued, he built walls and threw reckless tantrums,

Wars were fought and Earth was

BETRAYED

and held to ransom.

With great ability…
Comes great responsibility.

$E = mc^2$

In no time at all…

Man entered the age of abundance unlike anything seen before,

A world of extremes and paradoxes that demanded more and more….

He bent and exploited the natural world to his every whim and will,

Needs, wants and desires, *'click, click'* – empty baskets and voids to fill.

Man mined the rocks, farmed the wild and plundered the seas,

Yet consumption grew and spread like an infectious disease.

Man divided Earth and profiteered from her on an industrial scale.

She was

COMMODIFIED

into products, open for trade, ready for sale.

Over millions of years,
Earth buried the energy of life-forms in vaults deep in the ground,
What was once hidden and locked away, now discovered and found.

Soon after…
Man bored deep within her flesh with drills, cold and steel,
In agony, she bled barrels of black without a single

SQUEAL.

Man grew more advanced and found new ways of extraction,
He filled his containers to the brim, ready for exportation.

He jabbed her with needles and instruments of pain,
Fracturing the core of her being, as she quaked in vain.

Man feasted on a buffet of fossil remains year on year,

He gorged on the dark, he guzzled the dirty and fumed her atmosphere.

Man fuelled his industrial ambitions at a rate unprecedented,

In due course, Earth was devoured, depleted and devastated.

Man polluted the land, air, rivers and the seas,

Global warming ensued, then came a decline of the bees.

Man littered the seas and land with trash and plastic waste,

Degrading her soil and choking her oceans with earnest and haste.

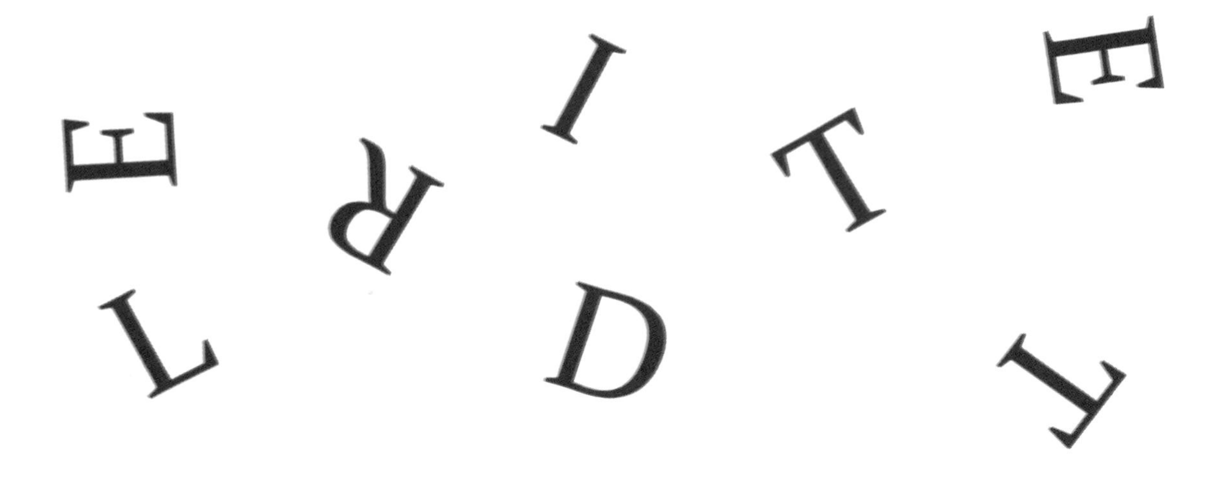

Moon watched on as animal numbers dwindled day after day,

'Surely Man can see, the collapse of biodiversity was well underway.

From the tiniest of insects to the largest mammals on land,

With great numbers disappearing – a catastrophe loomed close at hand!'

But still, Man did not change…

He hunted rhinos, tigers, sharks and whales… far and wide,

Once a species was targeted, there was nowhere on Earth it could hide.[1]

Man poached endangered species without care or consideration,

These remarkable forms life crafted, on the countdown to

EXTINCTION.

1 - A saying of David Attenborough

Man cleared grasslands and cut down trees at an unsettling rate,
Soon entire forests fell, animals looked on… awaiting their fate.

The loss of jungles and rainforests one after another,
Acted as a heat-trapping blanket, warming Earth even further.

Animals became bewildered by the ever-increasing heat,
Unable to bear it – they made a hasty retreat.
They abandoned their homes in sheer desperation,
As displaced refugees, they boarded the escalator to extinction.

And yet, still…
Man encroached further and deeper into the wild every day,
Whole habitats – desolated, countless more roamed

HOMELESS in utter dismay.

The planet continued to warm…

Soon the permafrost thawed and the Arctic ice began to recede,

The oceans simmered and the sea level rose at incredible speed.

Coastal wonderlands quickly faded, reefs bleached and turned white,

These wonders of colours, now devoid

WASTELANDS

– a truly haunting sight.

Marine life dwindled, instability ensued as ecosystems started to fail,

The impact had been far-reaching – from algae to the humpback whale.

Climate change affected the weather – now wild, volatile and extreme,

Frequent droughts, floods, hurricanes and everything in between.

Over the ages,

Moon watched as Man transformed the skies, the seas and the land,

His insatiable appetite to acquire new terrains, to grow and expand.

How Moon sighed and how he watched in dismay,

As the colours and hues of Earth were brushed over by shades of

GREY.

The impact of Man has been truly global and profound,
Earth rested on a tipping point, disaster loomed around....

On the horizon, a new existential threat eventually did appear,
Earth was facing collapse as the

HUMAN ASTEROID

drew near.

As Man burned brightly, Earth fell upon her knees.
Softly she pleaded, *'O child of mine, I am your mother... please...!'*

The Great Mother was silently crying,
Earth was being dismantled and slowly dying.

It was like a repeat – a déjà vu,
But what more could Moon do?

He watched from afar but was powerless to interfere,
All he could do was to offer a silent prayer.

He cried a thousand rivers until his tears dried up,
Deep scars and emotional craters became his bitter cup.

Place your bets now!

As Man continued to play

ROULETTE

with the future of Earth,
Moon was robbed of his youthful glow and celestial mirth.

As the devastation gathered increasingly in pace,
New chasms and canyons were etched upon Moon's noble face.

Yet…

Every day of every year, Moon smiled without fail,

He hid his sorrows behind an invisible veil.

In spite of his aching heart, he comforted Earth as much as he could muster.

Despite his broken spirit, he rose with full splendour and lustre.

He orbited Earth, radiating the warmth he received from the Sun.

He gave her strength and hope – she was the

UNIQUE ONE.

However,

Once every blue moon, he'd sigh deeply,

Overcome,

His sadness would overwhelm him completely.

His lament and heartache would be unveiled in full view,

Moon would rise sad and sombre in shades of

BLUE.

Then…

On one winter's day, something happened both remarkable and strange,

The Great Mother retaliated – *'Right, it's time for*

CHANGE!'

She gave birth to a virus, insignificantly small yet incredibly strong,

Swiftly it crossed land and seas - cities fell as it marched along.

Amidst the fear and chaos that ensued, Man's weakness was utterly exposed.

He retreated in haste and a mandatory self-exile was imposed!

Whilst he hid away…

Ships moored empty and the seas became calm and crystal clear,

Marine life flourished and dolphins swam freely without fear.

The forest echoed with wildlife as deforestation stopped,

The drilling ceased and pollution levels dropped.

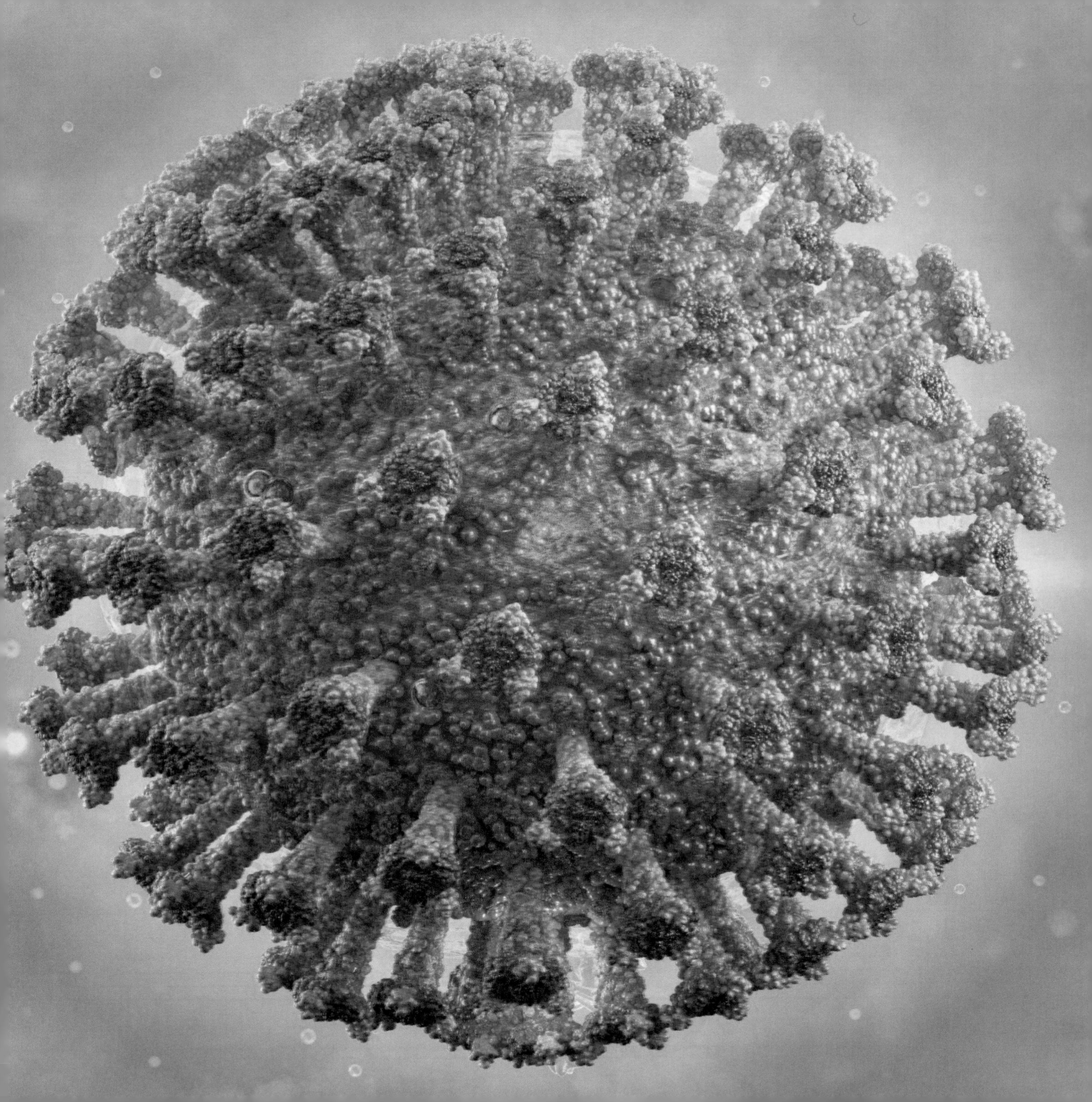

The trees blossomed, the birds returned and came to roost.
Animals roamed deserted streets as noise level reduced.

Trading closed, traffic dwindled and the air became crisp and clean,
Wars ended, planes grounded and the sky became
as blue as it's ever been.

During his exile and seclusion…
Man reflected deeply and saw the error of his self-serving ways,
Time to act and redress the wrongs of the old irresponsible days.

Man had the

GREAT REALISATION

– it was welcomed and long overdue.

Unsustainable practices cannot continue – it's time to begin anew.

In its place, a new idea emerged…

'Man is not separate from Earth, we're all one',

The Great Mother breathed relief – 'O *my son….'*

Moon thought:

'Earth is finally healing and change is on its way,

Man has learnt kindness and humility

– one can only hope and pray.'

And then, for the first time in a long, long while,

Moon remembered again – how to

SMILE."

**[Now, my story of the Earth, Moon and Man is all said and done.
Your friend, Ash-Shams – the Sun.]**

Printed in Great Britain
by Amazon